SpringerBriefs in Fire

Series editor

James A. Milke, College Park, MD, USA

For further volumes:
http://www.springer.com/series/10476

R. Thomas Long Jr. · Jason A. Sutula
Michael J. Kahn

Flammability of Cartoned Lithium Ion Batteries

 Springer

R. Thomas Long Jr.
Jason A. Sutula
Michael J. Kahn
Exponent, Inc.
Bowie, MD
USA

ISSN 2193-6595   ISSN 2193-6609 (electronic)
ISBN 978-1-4939-1076-2   ISBN 978-1-4939-1077-9   (eBook)
DOI 10.1007/978-1-4939-1077-9
Springer New York Heidelberg Dordrecht London

Library of Congress Control Number: 2014939537

Printed on acid-free paper

Springer is part of Springer Science+Business Media (www.springer.com)

Foreword

Lithium-ion battery cells and small battery packs (8–10 cells) are in wide consumer use today. Superior capacity has driven the demand for these batteries in electronic devices such as laptops, power tools, cameras, and cell phones. In 2011, the Foundation conducted a hazard and use assessment of these batteries, with a focus on developing information to inform fire protection strategies in storage. Since that time, the Foundation has conducted a survey of storage practices and developed a multiphase research strategy. This report presents the results of Phase II of the project, which is a comparative flammability characterization of common lithium-ion batteries to standard commodities in storage.

The content, opinions, and conclusions contained in this report are solely those of the authors.

Preface

This book gives a comparison of cartoned lithium-ion (Li-ion) batteries and FM Global standard commodities in a rack storage configuration. Subsequent large-scale fire tests were conducted by FM Global to assess the effectiveness of ceiling level only sprinkler protection. All data, test descriptions, data analysis, and figures in this report were graciously provided by FM Global. Exponent has relied on the FM Global testing report entitled, "Flammability Characterization of Lithium-ion Batteries in Bulk Storage," as a basis for this report [1].

This project was conducted in conjunction with the Property Insurance Research Group (PIRG) and was directed through FPRF.

Small format Li-ion battery commodities were selected to represent commercially available battery formats and Li-ion battery containing devices. The selected Li-ion battery types were individual 18,650 format cylindrical cells, power tool packs comprising 18,650 format cells, and polymer cells. The selected comparison commodities were the FM Global standard Class 2 and Cartoned Unexpanded Plastic (CUP). Two independent test series were conducted by FM Global. These tests represented a unique approach to hazard evaluation with limited commodity and were necessary due to the inordinate cost associated with Li-ion batteries.

The first test series evaluated the flammability characteristics of small format Li-ion batteries and the FM Global standard cartoned commodities in a three-tier high, single row, open frame, rack storage array. All tests conducted were free burn tests. In each test, cartoned commodity was only located in a portion of the rack where the initial fire growth was expected to lead to sprinkler operation (i.e., at the flues). The first test series was used to estimate the fire hazard from the cartoned commodity present at the time of the predicted first sprinkler operation.

A total of 13 tests were conducted. Ignition was achieved using an external fire. The key findings reported by FM Global included:

- The fire growth characteristics for the Li-ion batteries and the FM Global standard commodities that were evaluated exhibited similar fire development leading to the estimated time of first sprinkler operation.
- Commodity containing densely packed Li-ion batteries and minimal plastics (i.e., cylindrical and polymer cells) exhibited a delay in the battery involvement. For the Li-ion batteries used in this project, significant involvement was observed within 5 min after ignition.

- Commodity containing a significant quantity of loosely packed plastics (i.e., CUP and power tool packs) exhibited a rapid increase in the released energy due to plastics involvement early in the fire development. Battery involvement was not observable due to the contribution from the plastics.
- The CUP commodity exhibited a fire hazard leading to initial sprinkler operation that was similar or greater than the Li-ion battery products tested. Therefore, the CUP commodity was chosen as a suitable surrogate for Li-ion batteries in a bulk-packed rack storage test scenario, provided the fire protection system suppresses the fire prior to the time of significant Li-ion battery involvement.
- Without full-scale sprinklered testing experience with Li-ion batteries, protection system performance must preclude Li-ion battery involvement.

The second test series evaluated the level of protection provided by ceiling level only sprinklers. Two large-scale fire sprinkler tests were conducted with CUP commodity and were based on the reduced commodity testing approach data. Full-scale tests were conducted with CUP commodity due to the costs associated with purchasing Li-ion cells and/or battery packs.

In both large-scale fire tests, the CUP commodity cartons were breached by the fire before initial sprinkler operation, resulting in persistent, deep-seated flames beyond the predicted time of battery involvement. At this time, the adequacy of ceiling level sprinkler protection cannot be established without repeating the large-scale fire sprinkler tests using bulk-packed Li-ion cells and/or battery packs.

Reference

1. B. Ditch et al., *Flammability Characterization of Lithium-ion Batteries in Bulk Storage* (FM Global, Norwood, 2013)

Acknowledgments

The authors thank Benjamin Ditch and the entire FM Global Research Campus crew for their significant efforts in setting up, instrumenting, and conducting the fire and sprinkler tests and providing access to the data and analysis gathered during testing.

The authors further thank Kathleen Almand, Executive Director of the FPRF, the project technical panel, and the members of PIRG for the opportunity to work on this project. In addition, the funding provided by PIRG members and additional sponsors for the Li-ion batteries was essential to the economic feasibility of this project.

We also thank a number of our colleagues at Exponent who provided assistance, input, and advice.

Limitations

At the request of the Fire Protection Research Foundation (FPRF), Exponent has reported on the flammability characterization study of lithium-ion (Li-ion) batteries in bulk storage. This report summarizes a full-scale, reduced commodity fire testing comparison of cartoned Li-ion batteries and FM Global standard commodities in a rack storage configuration, as reported by FM Global. The scope of services performed during this assessment of the test data may not adequately address the needs of other users of this report, and any re-use of this report or its findings, conclusions, or recommendations presented herein are at the sole risk of the user.

The reduced-commodity approach of FM Global's large-scale tests and any recommendations made are strictly limited to the test conditions included in this report. The combined effects (including, but not limited to) of different storage heights, ceiling height, protection system design, battery density, state of charge, and battery type are yet to be fully understood and may not be inferred from these test results alone.

The findings formulated in this review are based on observations and information available at the time of writing. The findings presented herein are made to a reasonable degree of engineering certainty. If new data become available or there are perceived omissions or misstatements in this report, we ask that they are brought to our attention as soon as possible so that we have the opportunity to fully address them.

Limitations

Contents

Figures

Tables

Abstract

In 2011, FPRF conducted a hazard and use assessment of lithium-ion (Li-ion) batteries with a focus on bulk warehouse storage. FPRF has now completed the next phase (Phase IIB) of this program to develop the flammability characterization in an attempt to provide the basis for fire protection guidelines of common small format battery types in rack storage configurations. Based on the previous hazard assessment (Phase I) [1], and the storage survey (Phase IIA), this research entailed full-scale fire testing of three types of small format batteries: 18,650 format cylindrical Li-ion batteries, prismatic Li-ion polymer batteries of comparable capacity to the test 18,650 cells, and packaged power tool rechargeable battery packs with cylindrical cells.

Standard commodity classification testing typically involves three to four full-scale fire tests of eight or more pallet loads of the commodity. Due to the inordinate costs of acquiring pallets of Li-ion batteries, which may contain over 20,000 batteries, a modified approach was executed by FM Global. This modified approach benchmarked the flammability of a smaller quantity of cells/packs strategically arranged in a rack storage configuration against standard commodities in the same configuration and permitted testing of a reduced amount of commodity.

Li-ion chemistry has become the dominant rechargeable battery chemistry for consumer electronics. This chemistry is different from previously popular rechargeable battery chemistries (e.g., nickel metal hydride, nickel cadmium, and lead acid) in a number of ways. From a technological standpoint, because of high energy density, Li-ion technology has enabled or improved entire families of devices, such as laptops, power tools, cameras, and cell phones. The increased utilization of these devices has led to an influx of the bulk storage of Li-ion batteries and heightened the need for sprinkler protection options that address the hazards associated with Li-ion battery bulk storage fires.

Fire challenges associated with the bulk storage of Li-ion batteries are unique given the presence of a flammable organic electrolyte within the Li-ion battery as compared to the aqueous electrolytes typically found in other widely used battery types. When exposed to an external fire, Li-ion batteries can experience thermal runaway reactions resulting in the combustion of the flammable organics and the potential rupture of the battery [2]. NFPA 13, *Standard for the Installation of*

Sprinkler Systems [3], does not contain specific research-based sprinkler installation recommendations or protection requirements for Li-ion batteries.

This project was directed by FPRF. All resources associated with conducting the tests, as well as compiling the data and results, were generously donated by FM Global.

References

1. C. Mikolajczak et al., *Li-ion Batteries Hazard and Use Assessment* (Fire Protection Research Foundation, 2011), http://www.nfpa.org/assets/files/PDF/Research/RFLithiumIonBatteries Hazard.pdf
2. C. Arbizanni et al., Thermal stability and flammability of electrolytes for Li-ion batteries. *J. Power Sources* **196**(10), 4801–4805 (2011)
3. NFPA 13: Standard for the Installation of Sprinkler Systems, National Fire Protection Association (2010)

Chapter 1
Background

1.1 Project History

Phase I of this project concluded with a detailed report describing a hazard assessment of Li-ion batteries [1]. The key finding from Phase I was that the warehouse setting was frequent throughout the entire lifecycle of Li-ion batteries. In the warehouse setting, several failure modes of Li-ion batteries were identified, including mechanical abuse, electrical abuse, thermal abuse from an external fire, and internal fault. However, internal fault is unlikely unless the Li-ion cells are being actively charged while being stored. Thus, this failure mode is not directly applicable within a large storage warehouse setting and is outside the scope of the current project. Based on this, an external fire source attacking the stored commodities was selected as the ignition source for this test series (see Sect. 4.1). Several mitigation strategies for these failure modes are commonly employed, including reducing the state of charge (SOC) of stored cells (i.e., 50 % SOC or less), placing cells in packaging designed to prevent mechanical and external short circuit damage, and the use of various fire protection systems.

Phase II extended the work completed in Phase I with the ultimate goal of establishing specific fire protection guidance for bulk warehouse storage of small format Li-ion batteries. Phase II was segmented into two components: Phase IIA and Phase IIB. Phase IIA consisted of a survey of common Li-ion batteries and storage conditions found in warehouse storage settings. The survey gathered data from groups that store batteries, cells, or devices containing batteries or cells. The primary responders included groups from manufacturing, research, and recycling.

Based on a summary of the responses, several storage details related to Li-ion batteries were identified. First, the most common form factor was a small cylindrical cell. Second, almost all of the response groups were engaged in storage of Li-ion cells or batteries. Third, the Li-ion batteries or cells were typically packaged in cardboard boxes. The boxes were commonly on wood pallets and were encapsulated. Finally, the palletized loads were stored in a rack storage configuration. Movable racks were more commonly found than fixed racks, and the shelves were likely to be perforated.

R. T. Long Jr. et al., *Flammability of Cartoned Lithium Ion Batteries*,
SpringerBriefs in Fire, DOI: 10.1007/978-1-4939-1077-9_1,
© Fire Protection Research Foundation 2014

Based on the range of Li-ion cell types, 18,650 format cylindrical Li-ion batteries, prismatic Li-ion polymer batteries of comparable capacity to the test 18,650 cells, and packaged power tool rechargeable battery packs with cylindrical cells were identified as the most pertinent for the analysis. These batteries are typically found in a host of different commodities, including, portable GPS devices, portable game players, portable DVD players, portable televisions, portable radios, cellular phones, music players, electronic readers, notebook computers, cordless headphones, universal remote controls, cameras, camcorders, two-way radios, rechargeable vacuums, electric razors, electric toothbrushes, and electric vehicles.

In evaluating the use of various fire protection systems, one of the main knowledge gaps identified in Phase I was the lack of research-based sprinkler protection guidance for storage of Li-ion batteries. Water based automatic sprinkler systems are commonly used in warehouses, therefore, water based suppression was chosen as a starting point for evaluating fire protection strategies for Li-ion batteries. At present, there is no designated fire protection suppression strategy for bulk packaged Li-ion cells, larger format Li-ion cells, or Li-ion cells contained in or packed with other equipment. NFPA 13 does not provide a specific recommendation for the protection of Li-ion cells or complete batteries, and it is not known if water is the most appropriate extinguishing medium for Li-ion batteries.

Phase IIB included full-scale fire testing and is described in detail in the remainder of this report.

1.2 Cost Feasibility of the FM Global Standard Method

The FM Global standard method used to evaluate protection requirements is to classify bulk packaged materials for fire sprinkler protection through "commodity classification" [2, 3]. The FM Global standard method was not feasible for this project due to the excessive cost associated with acquiring a sufficient quantity of Li-ion batteries. For example, the 18,650-format cylindrical cells are commonly packaged in corrugated board boxes that contained 200 cells per box and 96 full boxes per pallet. This equates to 19,200 cylindrical cells per pallet at a cost of approximately $60,000 US dollars per pallet load. To conduct a single three rack high test would require a minimum of six pallets stacked vertically with a flu space between. Thus, an innovative approach was required to permit testing of a reduced amount of Li-ion commodity.

It is important to note that the reduced commodity approach does not provide the same level of information regarding protection system performance gained through commodity classification of sprinklered large-scale tests using bulk-stored Li-ion batteries. Consequently, a two-part test approach was necessary to provide general protection guidance for Li-ion batteries, which incorporated an assessment of the flammability characteristics of the test commodity and an independent estimate of the effectiveness of sprinkler protection.

References

1. C. Mikolajczak et al., Li-ion batteries hazard and use assessment. Fire Protection Research Foundation (2011), http://www.nfpa.org/assets/files/PDF/Research/RFLithiumIonBatteries Hazard.pdf
2. Y. Xin et al., Assessment of commodity classification for sprinkler protection using representative fuels. J. Fire Saf. Sci. **9**, 527–538 (2008)
3. Y. Xin, Storage height effects on fire growth rates of cartoned commodities, in *Seventh International Seminar on Fire and Explosion Hazards*, Providence, 2013

Chapter 2
Commodity Description

2.1 FM Global Standard Commodities

Two FM Global standard commodities were selected for this project: Class 2 and Cartoned Unexpanded Plastic (CUP). The Class 2 commodity consists of three double-walled corrugated paper cartons. Inside the cartons is a five-sided sheet metal liner, representing non-combustible content and the cartoned liner is supported on an ordinary, two-way, slatted deck, hardwood pallet (see Fig. 2.1).

The FM Global standard CUP commodity consists of Group A unexpanded rigid crystalline polystyrene cups packaged in single-walled, corrugated paper cartons. Cups are individually compartmentalized with corrugated paper partitions and are arranged in five layers (see Fig. 2.2).

2.2 Li-ion 18,650 Cylindrical Cells

The Li-ion 18,650-format cylindrical cell has approximate dimensions of 0.7 inches in diameter and 2.6 inches long. The cell is constructed by winding long strips of electrodes into a "jelly roll" configuration, which is then inserted into a hard metal case and sealed with gaskets (see Fig. 2.3). The packaging, as received from the manufacturer, consisted of two inner corrugated board cartons within an outer corrugated board carton. Each inner carton contained 100 cylindrical cells separated by paperboard partitions (see Fig. 2.4).

2.3 Li-ion 18 V Power Tool Packs

The power tool packs tested were comprised of ten 18,650-format cylindrical cells encased in a plastic case. The individual battery packs were encased in plastic blister packs for display, as shown in Fig. 2.5. The power tool packs are comprised

R. T. Long Jr. et al., *Flammability of Cartoned Lithium Ion Batteries*,
SpringerBriefs in Fire, DOI: 10.1007/978-1-4939-1077-9_2,
© Fire Protection Research Foundation 2014

Fig. 2.1 FM Global standard
class 2 commodity (courtesy
of FM Global)

Fig. 2.2 FM Global standard
CUP commodity (courtesy of
FM Global)

Fig. 2.3 Li-ion 18,650-
format cylindrical cell
(courtesy of FM Global)

Fig. 2.4 Li-ion cylindrical
cell packaging as received
from manufacturer (courtesy
of FM Global)

Fig. 2.5 Li-ion power tool
pack shown with blister case
(courtesy of FM Global)

Fig. 2.6 Li-ion power tool
packs in cartons on a pallet
(courtesy of FM Global)

of significant quantities of plastics (i.e., the battery case itself and the clear blister packaging), unlike the cylindrical and polymer cells. Figure 2.6 depicts the Li-ion 18 V power tool packs in cartons on a pallet. Each carton contained four power tool packs.

2.4 Li-ion Polymer Cells

The Li-ion polymer cells are enclosed in a soft-case to reduce the overall size and weight and have approximate dimensions of 3.9 inches in length, 1.6 inches in width, and 0.23 inches in thickness. The cell is constructed by winding long strips of electrodes into a "jelly roll" configuration, which is then enclosed in a polymer-coated aluminum pouch with heat-sealed seams (see Fig. 2.7). There were 144 cells in each carton, as shown in Fig. 2.8. Each of those cartons contained two smaller cartons of 72 cells.

Fig. 2.7 Li-ion polymer cell
(courtesy of FM Global)

Fig. 2.8 Li-ion polymer
cells in cartons on a pallet
(courtesy of FM Global)

Table 2.1 Summary of battery cell characteristics

Parameter	Power tool 18,650	18,650	Li-polymer
Nominal voltage	3.7 V	3.7 V	3.7 V
Nominal capacity	1,300 mAh	2,600 mAh	2,700 mAh
Mass of cell	42.9 g	47.2 g	50.0 g
Approximate mass of electrolyte solvent	3.3 g	2.6 g	4.0 g
Cell chemistry	Lithium nickel maganese cobalt oxide (NMC)	Lithium cobalt oxide (LCO)	Lithium cobalt oxide (LCO)
Approximate state of charge (SOC) as received	50%	40%	60%

2.5 Li-ion Battery Cell Characteristics

A summary of the characteristics of battery cells tested is provided in Table 2.1.
   Refer to Sect. 4.3 of the FM Global report [1] for further detail on information presented in this section.

Reference

1. B. Ditch et al., Flammability characterization of lithium-ion batteries in bulk storage. FM
   Global, March 2013

Chapter 3
FM Global Reduced-Commodity Testing

All testing descriptions and data included in this section were extracted from the FM Global report [1].

3.1 Testing Configuration and Setup

A reduced-commodity test was developed by FM Global that captured the flammability characteristics inherent in a rack storage fire while limiting the total quantity of test commodity to approximately one pallet load per test due to the significant costs, as previously discussed.

Thirteen reduced-commodity fire tests were conducted focusing on a comparison of the flammability characteristics of FM Global standard commodities and the three types of cartoned Li-ion battery products. All tests were free burn fire tests that measured the heat release rate and the evaluated time of battery involvement for the Li-ion products. Each test was conducted in a two by one by three high pallet load rack storage arrangement that represents storage up to 15 ft tall.

The array consisted of a three tier high, open-frame, single-row steel rack. In each test, only the ignition flue area of the array was lined with commodity. The bottom tier of the array was comprised of a non-combustible product (i.e., metal liner) on a wood pallet. The upper tiers consisted of the same non-combustible product lined with test commodity on the flue faces (see Fig. 3.1). Table 3.1 presents the total amount of reduced-commodity that was utilized in each test separated by tier.

Ignition was achieved with a propane ring burner centered in the transverse flue below the second tier test commodity. The heat release rate of the burner was approximately 45 kW.

Documentation for each test included video, infrared (IR) video, and still photography. Carbon dioxide (CO_2), carbon monoxide (CO), total hydrocarbons (THC), and depletion of oxygen (O_2) were measured in the exhaust. The propane gas flow rate to the burner and the mass loss of the commodity were also recorded.

R. T. Long Jr. et al., *Flammability of Cartoned Lithium Ion Batteries*,
SpringerBriefs in Fire, DOI: 10.1007/978-1-4939-1077-9_3,
© Fire Protection Research Foundation 2014

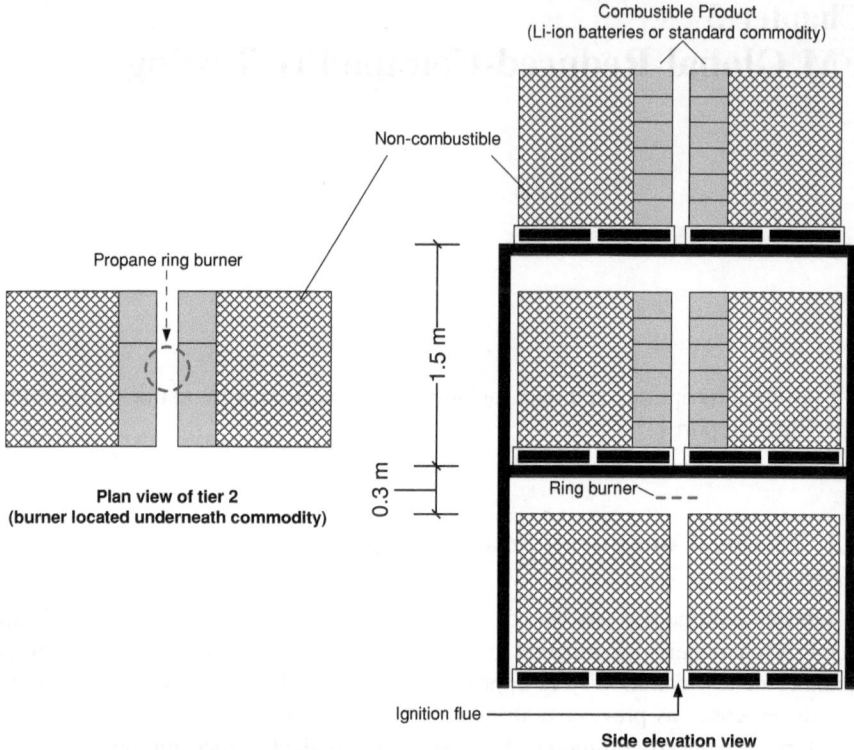

Fig. 3.1 Schematic of reduced-commodity rack storage test; plan view of tier 2 (*left*) and side elevation view (*right*); (courtesy of FM Global)

Table 3.1 Amount of commodity type utilized, separated by tier

Reduced commodity type	Tier 1 amount	Tier 2 amount	Tier 3 amount
18,650-format cylindrical	None	9,600 cylindrical cells	9,600 cylindrical cells
Power tool pack	None	100 packs	100 packs
Polymer	None	7,776 cells	7,776 cells

Thermocouples were used to monitor the internal heating of the commodity during the fire tests (see Fig. 3.2).

3.2 Testing Results

The reduced-commodity fire test results compared the flammability characteristics of the standard commodities to the Li-ion battery products in a consistent geometry, however, with limited commodities located only at the flue faces. The

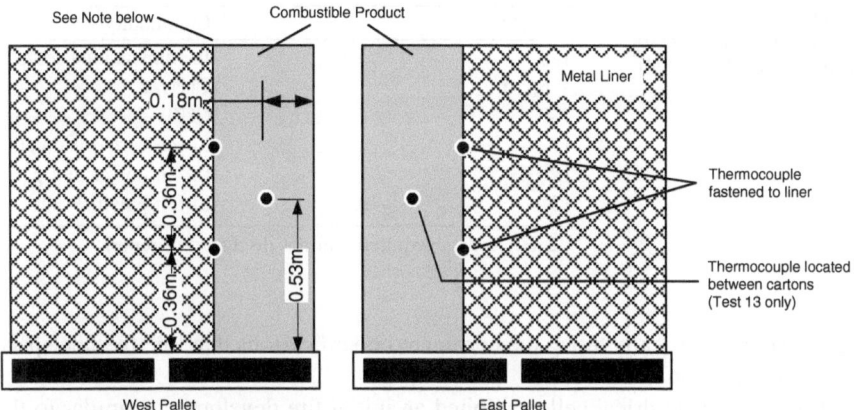

Fig. 3.2 Thermocouple locations for test pallets (courtesy of FM Global). *Note* Interface of combustible product and metal liner varies for each commodity

commodities were subjected to an external ignition source that first ignited the cartoned packaging and then subsequently the Li-ion battery products. Table 3.2 provides an overview of the 13 tests.

Comparison of the flammability characteristics between the standard commodities and Li-ion batteries can only occur during the period where the fire did not propagate beyond the commodity of interest. Once the fire reaches the extent of the combustible commodity, the results can no longer be used to evaluate sprinkler response, since further fire propagation is not feasible. It is important to note that additional information regarding the overall fire hazard of each commodity can be obtained after the period of flammability characterization, in particular, the time of significant battery involvement.

The four monitoring techniques utilized to evaluate propagation penetration in the commodity included standard video cameras to monitor the location of flame attachment; infrared imaging cameras to monitor external heating of the commodity; thermocouples attached to the commodity-metal liner interface to monitor internal heating of the commodity; and product collapse. The FM Global standard commodities did not exhibit collapse before termination of the fire test and are not discussed.

The four monitoring techniques described above provide data that was assessed to determine the nominal period of flammability characterization. Data from each of the techniques was compiled and was found to extend to approximately 75 seconds after ignition. Table 3.3 shows the period of flammability characterization based on the monitoring techniques. The analysis method for each technique is described in further detail in Sect. 4.7 of the FM Global testing report [1].

The convective heat release rates for each commodity are shown in Fig. 3.3. As seen in this figure, each commodity exhibited a similar initial fire growth as the

Table 3.2 Reduced-commodity test summary (courtesy of FM Global)

Test [#]	Commodity
1–3	Instrumentation setup[a]
4–6; 10	Class 2
7–9	CUP
11	Li-ion cylindrical cells
12	Li-ion power tool packs
13	Li-ion polymer cells

[a] Limited data acquired and not discussed in report

flames spread vertically along the corrugated board cartons that line the flue space above the ignition zone.

The Li-ion cylindrical cells exhibited an initial fire development similar to the Class 2 commodity until 110 s, when the fire size began to decline due to consumption of the carton material. The fire size steadily declined until 200 s, when the Li-ion cells became significantly involved and the fire reached a maximum of 3,900 kW at 690 seconds. The increases in fire size at 490 and 685 s were a result of product collapse.

The Li-ion power tool packs exhibited a fire development trend similar to the CUP commodity. A delay in fire growth occurred at 55 s as the flames penetrated the cartons and the plastic components of the battery packs became involved in the fire. This delay was observed as a temporary plateau in the heat release curve from 55 to 75 s. The fire then grew steadily from 75 s until the fire reached a maximum of 1,900 kW. The fire size then decreased to approximately 1,250 kW, remained steady, and then decreased as a majority of the combustibles were consumed.

The Li-ion polymer cells exhibited an initial fire development similar to the Class 2 commodity until 110 s when the fire size began to decline due to the consumption of the cartoned material. The fire size steadily declined until 305 s when the Li-ion cells became significantly involved, and the fire reached a maximum of 3,800 kW at 750 s.

The heat release rates measurements from the initial vertical fire spread support the assumption that all of the tested cartoned commodities exhibit similar initial fire development in a three-tier high rack storage array. The subsequent breach of the cartons highlights the impact of the stored contents of the cartoned commodities. For cartons containing significant quantities of loosely packed plastics (i.e., CUP and power tool packs), involvement of the plastic resulted in a rapid increase in the heat release early in the fire development. For cartoned commodities containing densely packed Li-ion batteries and lesser amounts of plastics (i.e., Li-ion cylindrical and polymer cells), the fire growth was delayed until the Li-ion batteries became significantly involved in a three tier high rack storage array.

Significant battery involvement was qualitatively assessed based on visual observations of the fire and a simultaneous increase in the convective heat release rate beyond a threshold value. Figure 3.4 shows the convective heat release rates

Table 3.3 Period of flammability characterization using all indicators; data presented as time (s) after ignition (courtesy of FM Global)

Commodity	Flame attachment	External heating	Internal heating	Product collapse
Class 2	60–90	60–90	n/a	none
CUP	60–90	60–90	160	none
Li-ion cylindrical cells	60–90	60–90	310	497
Li-ion power tool packs	60–90	60–90	120	94
Li-ion polymer cells	60–90	60–90	315	540

Fig. 3.3 Convective heat release rates for FM Global standard commodities and Li-ion battery commodities (courtesy of FM Global)

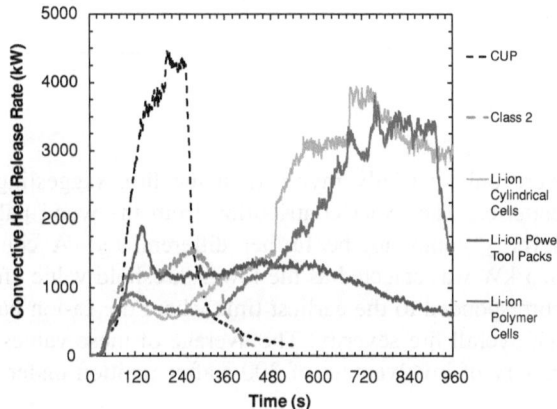

associated with the Class 2 commodity, Li-ion power tool packs, Li-ion cylindrical cells, and Li-ion polymer cells.

The time of battery involvement was evaluated by FM Global by examining the early fire growth. During the Class 2 commodity and Li-ion power tool pack tests, the majority of the test commodity was consumed during the initial fire growth. The subsequent nominal steady-state heat release rate of 1,250 kW was a result of the combustion of the wood pallets on the second and third tier (1,250 kW/4 pallets is equal to 312.5 kW/pallet). The contribution from the wood pallets to the heat release was assumed to be the same for all tests. Any increase in the heat release rate can then be attributed to the combustion of the Li-ion batteries. 1,250 kW was selected as the upper threshold value corresponding to the latest time where the Li-ion battery products did not significantly contribute to the fire. The 1,250 kW threshold was exceeded at 385 and 470 s, for the Li-ion cylindrical cells and polymer cells, respectively. No estimate was made for the power tool packs due to the significant quantities of plastics.

The time of battery involvement was also estimated by FM Global earlier in the fire development. During the Li-ion cylindrical cell and polymer cell tests, the convective heat release rate steadily decreased after the initial vertical flame spread along the cartoned commodity packaging to approximately 600 kW at 200 and 305 s, respectively. The heat release rate then steadily increased to the upper threshold value of 1,250 kW. In both tests, the wood pallets on the second tier

Fig. 3.4 Convective HRR
for Class 2 commodity and
Li-ion battery commodities
(courtesy of FM Global)

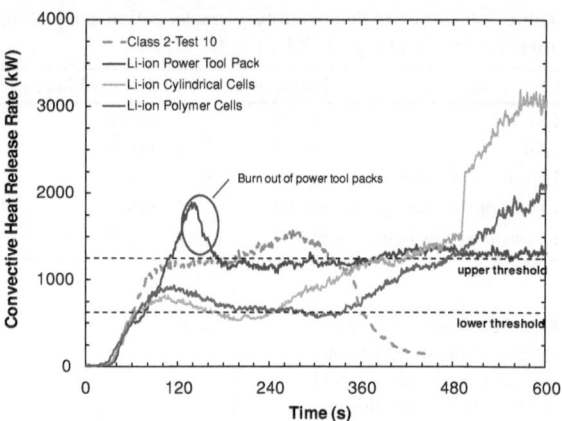

were only partially involved in the fire, suggesting involvement of the Li-ion batteries. The exact contribution from the wood pallets and batteries to the fire severity could not be further differentiated. A convective heat release rate of 625 kW was selected as the lower threshold value after the initial fire growth that corresponded to the earliest time where the Li-ion battery products contributed to the overall fire severity. The average of these values results in a nominal time of battery of involvement of 300 s after ignition under free-burn fire conditions.

3.3 Predicted Sprinkler Response

A theoretical method to calculate the response time of sprinkler links to rack storage fires was used to predict the sprinkler operation time [2–5]. Table 3.4 presents the predicted quick-response link operation times for all commodities evaluated, assuming 25-foot and 30-foot ceiling heights. The general agreement of the fire growth characteristics supports the assumption that the cartoned commodities exhibited similar fire development leading to quick response sprinkler operation in a three tier high rack storage array. The predicted link operation time of 87 s for the Li-ion power tool packs assuming a quick response sprinkler under a 30-foot ceiling occurred after the flammability characterization period of approximately 75 s (see Sect. 3.2). Therefore, the reduced commodity array of power tool packs did not contain sufficient test commodity to evaluate the performance of quick response sprinklers under a 30-foot ceiling.

Table 3.5 presents the predicted standard response link operation times for all commodities included in this project, assuming 25-foot and 30-foot ceiling heights. For a 25-foot ceiling, the general agreement of the fire growth characteristics support the assumption that the cartoned commodities exhibited similar fire development leading to standard response sprinkler operation in a three tier high rack storage array. For a 30-foot ceiling, the standard response sprinkler link

Table 3.4 Predicted quick response sprinkler link operation time and corresponding fire growth characteristics (courtesy of FM Global)

Ceiling height	7.6 m (25 ft)			9.1 m (30 ft)		
Commodity	Link operation (s)	Q_{be}^a (kW)	Fire growth (kW/s)	Link operation (S)	Q_{be}^a (kW)	Fire growth (kW/s)
Class 2[b]	59	209	15	65	367	24
CUP[c]	43	232	16	52	321	11
Li-ion Cylindrical Cells	44	284	23	76	405	23
Li-ion Power Tool Packs	51	282	25	87	426	29
Li-ion Polymer Cells	41	256	16	77	370	13

[a] Fire size at the time of the predicted sprinkler operation. In a commodity classification test this coincides with the start of water application
[b] Average values for Tests 4–6, 10
[c] Average values for Tests 7–9

Table 3.5 Predicted standard response sprinkler link operation time and corresponding fire growth characteristics (courtesy of FM Global)

Ceiling height	7.6 m (25 ft)			9.1 m (30 ft)		
Commodity	Link operation (s)	Q_{be}^a (kW)	Fire growth (kW/s)	Link operation (s)	Q_{be}^a (kW)	Fire growth (kW/s)
Class 2[b]	77	603	22	90	799	14
CUP[c]	70	431	7[d]	86	818	30
Li-ion cylindrical cells	62	577	12	256	699	4
Li-ion power tool packs	70	497	−1[d]	125	719	16
Li-ion polymer cells	64	553	11	144	782	9

[a] Fire size at the time of the predicted sprinkler operation. In a commodity classification test this coincides with the start of water application
[b] Average values for Tests 4-6, 10
[c] Average values for Tests 7–9
[d] Small and negative fire growth rate due to a plateau in the growth of the fire during the transition from the outer carton to the stored plastic

operation times widely ranged from 86 seconds to 256 s, which are beyond the flammability characterization period of approximately 75 s (see Sect. 3.2). Therefore, the reduced commodity array of all cartoned commodities tested did not contain a sufficient quantity of test commodity to evaluate the performance of standard response sprinklers under a 30-foot ceiling.

Refer to Sect. 4 of the FM Global report [1] for further detail on information presented in this section.

References

1. B. Ditch et al., Flammability characterization of lithium-ion batteries in bulk storage. FM Global, March 2013
2. G. Heskestad, Investigation of a new sprinkler sensitivity approval test: the plunge test. FMRC Technical Report, Serial No. 22485, RC 76-T-50, 1976
3. G. Heskestad, Pressure profiles generated by fire plumes impacting on horizontal ceilings. FMRC Technical Report, 0F0E1.RU, August 1980
4. H.Z. You et al., Strong buoyant plumes of growing rack storage fires, in *Twentieth Symposium (International) on Combustion*, The Combustion Institute, pp. 1547–1554, 1984
5. G. Heskestad et al., Plung test for determination of sprinkler sensitivity. FMRC Technical Report, J.I 3A1E2.RR, December 1980

Chapter 4
FM Global Large-Scale Fire Sprinkler Testing

4.1 Test Configuration

Two large-scale fire tests were conducted to evaluate the level of protection provided by a ceiling only sprinkler system. The performance objective was for the ceiling only sprinkler system to extinguish a large-scale rack storage fire of CUP commodity. The passing criteria for the test was that the fire must be nearly extinguished before the predicted time of significant battery involvement established for the Li-ion battery commodities in the reduced-commodity tests (see Sect. 5.1).

Test 14 and Test 15 were conducted using a three pallet high load configuration on an open-frame, double-row steel rack. This array size represents storage up to 15 ft high. The ceiling heights were 30 and 25 ft above the floor for Test 14 and Test 15, respectively.

Sprinkler protection for each test was provided by FM approved pendent-type sprinklers with a 165°F temperature rating. For Test 14, the sprinklers had a k-factor of 25.2 gpm/psi$^{1/2}$ and were installed 18 inches below the ceiling. A nominal operating pressure of 25 psig provided a discharge of 125 gpm per sprinkler. The sprinklers were installed on 10-foot by 10-foot spacing, resulting in a significant density of 1.25 gpm/ft^2.

For Test 15, the sprinklers had a k-factor of 14 gpm/psi$^{1/2}$ and were installed 14 inches below the ceiling. A nominal operating pressure of 75 psig provided a discharge of 120 gpm per sprinkler. The sprinklers were installed on 10-foot by 10-foot spacing, resulting in a significant density of 1.2 gpm/ft^2.

Ignition was achieved with two FM Global standard half igniters. The overall test configuration is shown in Fig. 4.1. A single-row target array contained two pallet loads of the Class 2 commodity across a 4-foot aisle to the west of the main array to improve viewing angles of the ignition area.

Documentation for each test included video, IR and still photography, temperature measurements, flow and pressure measurements for the sprinkler system, and exhaust gas measurements of carbon dioxide, carbon monoxide, total hydrocarbons, and the depletion of oxygen. Supplementary instrumentation was added

R. T. Long Jr. et al., *Flammability of Cartoned Lithium Ion Batteries*,
SpringerBriefs in Fire, DOI: 10.1007/978-1-4939-1077-9_4,
© Fire Protection Research Foundation 2014

Fig. 4.1 Overall test
configuration for large-scale
validation tests, plan view
(courtesy of FM Global)

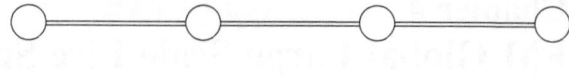

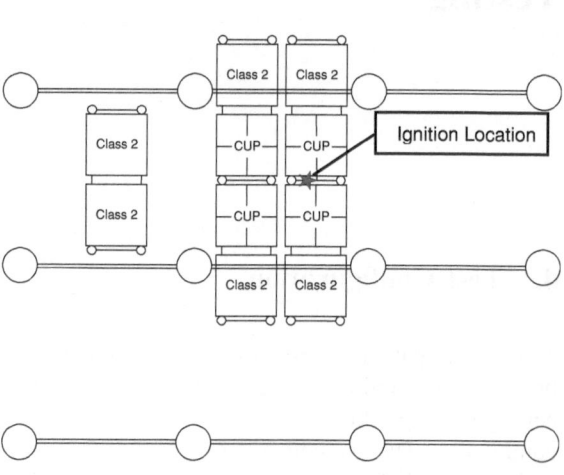

to assess the heating potential (magnitude and duration) exhibited by the burning
combustibles during the fire in the form of sixteen thermocouples located on the
inside of the commodity on the vertical side of the carton facing the ignition flue
and four thermocouples located on the horizontal top side of the carton. These
thermocouples assessed the presence of flames and the evaluation criterion was
based on the reduced commodity test results (see Sect. 5.7 of the FM Global report
[1] for further details).

4.2 Sprinkler Effectiveness Criteria

The primary evaluation criteria used for evaluating sprinkler effectiveness was the
number of sprinklers that operated, the extent of fire damage, heating potential for
material within the commodity cartons, and the magnitude and duration of ceiling
steel temperatures.

4.3 Fire Sprinkler Test Results

Test 14: A total of two sprinklers operated, with the first operation occurring at
1 min 48 s, followed by the second at 2 min 1 s after ignition. The overall fire
spread remained within the confines of the test array; however, damage to the

Table 4.1 Summary of large-scale tests (courtesy of FM Global)

Configuration and results	Test 14	Test 15
Test configuration		
Commodity	CUP, double row rack	
Commodity/ceiling height [m (ft)]	4.6/9.1 (15/30)	4.6/7.6 (15/25)
Main array located below—number of sprinklers	4	
Test results		
Sprinklers operations	2	4
Total energy released[a] [MJ (BTU $\times 10^3$)]	2, $\pm$ 360 (1,900 $\pm$ 340)	620 $\pm$ 25 (590 $\pm$ 120)
Consumed CUP commodity [pallet load equivalent]	1.5	0.5
Target jump (west only) @ Time [min:s]	None	None
Maximum one-minute steel temperature [°C (°F)] @ Time min:s]	37 (99) @ 5:09	41 (110) @ 1:48
Test termination [min:s]	20:00	20:00

[a] Based on generation rates of CO and CO_2

commodity surrounding the ignition zone and thermocouple measurements within the commodity both exceeded the evaluation criteria. A summary of the test results is provided in Table 4.1.

Test 15: a total of four sprinklers operated between 1 minute 38 s and 1 min 41 s after ignition. The overall fire spread was reduced compared to Test 14; however, persistent flames were observed throughout the test duration and thermocouple measurements within the commodity exceeded the evaluation criteria. A summary of the test results is provided in Table 4.1.

Refer to Sect. 5 of the FM Global report [1] for further detail on information presented in this section.

Reference

1. B. Ditch et al., Flammability characterization of lithium-ion batteries in bulk storage. FM Global, March 2013

Chapter 5
Conclusions

A total of 13 tests were conducted. Ignition was achieved using an external fire. The key findings reported by FM Global [1] included:

- The fire growth characteristics for the Li-ion batteries and the FM Global standard commodities that were evaluated exhibited similar fire development leading to the estimated time of first sprinkler operation.
- Commodity containing densely packed Li-ion batteries and minimal plastics (i.e., cylindrical and polymer cells) exhibited a delay in the battery involvement. For the Li-ion batteries used in this project, significant involvement was observed within five minutes after ignition.
- Commodity containing a significant quantity of loosely packed plastics (i.e., CUP and power tool packs) exhibited a rapid increase in the released energy due to plastics involvement early in the fire development. Battery involvement was not observable due to the contribution from the plastics.
- The CUP commodity exhibited a fire hazard leading to initial sprinkler operation that was similar or greater than the Li-ion battery products tested. Therefore, the CUP commodity was chosen as a suitable surrogate for Li-ion batteries in a bulk packed rack storage test scenario, provided the fire protection system suppresses the fire prior to the time of significant Li-ion battery involvement.
- Without full-scale sprinklered testing experience with Li-ion batteries, protection system performance must preclude Li-ion battery involvement.

The primary objective of this project was to develop a flammability characterization of Li-ion batteries that may serve as the basis for fire protection guidelines of common battery types in rack storage configurations. The large amount of fire and flammability characterization data collected for the three battery commodities tested provides a significant source of information on a high interest subject that was previously populated with little, if any, existing large-scale fire and flammability test data. While water application tests were not part of the original scope, two full-scale sprinklered tests were added to the protocol late in the project.

R. T. Long Jr. et al., *Flammability of Cartoned Lithium Ion Batteries*,
SpringerBriefs in Fire, DOI: 10.1007/978-1-4939-1077-9_5,
© Fire Protection Research Foundation 2014

Water based fire protection recommendations for small format bulk packaged Li-ion batteries in rack storage configurations could not be derived directly from the results of the reduced commodity full-scale Li-ion battery tests or through the full-scale CUP commodity ceiling only sprinkler tests as conducted by FM Global. Based on the results of the reduced commodity and full-scale tests conducted by FM Global, further engineering analysis and guidance is necessary to support specific sprinkler protection criteria for bulk packaged cartoned Li-ion batteries in rack storage configurations.

The FM Global report includes specific protection recommendations based on historical FM protection requirements for other commodities that exhibit similar hazard characteristics to Li-ion batteries. This information is included in Sect. 8, *Recommendations,* of the FM Global report [1].

Reference

1. B. Ditch et al., Flammability characterization of lithium-ion batteries in bulk storage. FM Global, March 2013

Chapter 6
Possible Future Work

The following possible future work is suggested (Phase IIC) to further understand protection requirements for bulk packaged small format Li-ion batteries in rack storage configurations:

- Determine the overall effectiveness of reduced-scale testing through additional reduced commodity tests utilizing power tool packs and/or possibly including additional small format Li-ion batteries with appreciable amounts of plastics, such as portable computer batteries;
- Determine the effectiveness of reduced-scale testing through additional reduced commodity tests using hard case bulk packaged prismatic format cells;
- Determine the effectiveness of reduced-scale testing through additional reduced commodity tests using an internal cell fault scenario;
- Develop, as possible, both reduced-scale and full-scale testing protocols based on the results of additional reduced commodity tests;
- Re-evaluate single credible worst-case commodity for testing based on any additional reduced-scale testing conducted;
- Develop and execute a full-scale testing protocol utilizing water as the extinguishing agent for full-scale testing of the chosen single credible worst case Li-ion battery commodity. Use the testing protocol to evaluate both ceiling only and in-rack protection strategies for various storage heights and arrangements and to include an assessment of sprinkler system and firefighting runoff water impact on the environment, as well as flaming projectiles and any potential re-ignition scenarios; and
- Make final protection system recommendations and fire service overhaul guidelines.

R. T. Long Jr. et al., *Flammability of Cartoned Lithium Ion Batteries*,
SpringerBriefs in Fire, DOI: 10.1007/978-1-4939-1077-9_6,
© Fire Protection Research Foundation 2014

Notes

Notes

Notes

Notes

Notes

Notes

Notes

Notes